Utilizing Microsoft Teams as a Modern Method of Intelligent Communications

Mini-Book Technology Series – Book 5

Authors:
Rand Morimoto, Ph.D.
David Ross, MCITP

DEDICATION

I dedicate this book to my children: Noble, Kelly, Chip, and Eduardo, may you always strive to try your best!
-- Rand

I dedicate this book to Annette – thanks for lighting up the room with your smile and enriching the lives of your three Ross men.
-- David

TABLE OF CONTENTS

ABSTRACT

Microsoft Teams is a cloud-based platform that Microsoft created to combine several standalone modes of communications into what Microsoft calls Intelligent Communications. Teams combines in a single interface: persistent chat, Web conferencing, video conferencing, person to person voice, cloud PBX telephony, file sharing, content collaboration, group calendaring, and team-focused communications. This book covers the features and functions built-in to Microsoft Teams, and more importantly shares best practices how organizations knit together the capabilities in Teams that they can then leverage to improve communications both internal and external to their enterprise.

INTRODUCTION

1 WHAT IS MICROSOFT TEAMS

Over the years, content collaboration and enterprise communications has shifted from old fashion cork boards with bulletins pinned on them, to digital bulletin boards, to attachments on emails, to Intranet sites and SharePoint sites, to a variety of SaaS-based group chat and collaboration tools (like Slack, HipChat, and the like).

Microsoft released "Microsoft Teams" as a tool to consolidate, replace, and update legacy uses of SharePoint, Skype, and 3rd party products, however organizations frequently ask "what can we do with Microsoft Teams in our enterprise, and what are the best practices utilized by other organizations in getting the most out of their Microsoft Teams implementation?"

First of all, Microsoft Teams does a LOT, from file sharing, to Web conferencing, to enabling team collaboration, and most organizations find they "use it all." The balance of this chapter and the overall content of this books helps provide a framework and structure around all that Teams does and the best practices leveraged by organizations to gain user adoption and enable team collaboration and intelligent communications in the enterprise.

Microsoft Teams for Enterprise Communications and Collaboration

With the current feature sets that are built in to Microsoft Teams, organizations are able to consolidate several tools currently used in the organization into just using the Teams platform to provide content sharing and collaboration in a unified environment. Specifically:

- Universal Client - Microsoft Teams provides a universal client that is the same/identical across all business platforms including Microsoft Windows, Apple Mac, iOS, Android, and Web. This means participants in Teams can work from any platform and all platforms independent of what "system" they are working on at the time.

- Teams Landing Page - When users log on to Teams, they see all of the teams they are members of, can Favorite key teams "channels," and actively work on critical content. A sample landing page for users of Microsoft Teams can be seen in Figure 1-1.

Figure 1-1: Sample Microsoft Teams Landing Page

- Files Access - Teams consolidates SharePoint document libraries and allows users to access files either through the Teams app or Web interface (as shown in Figure 1-2), or for those more

10

comfortable using Windows Explorer and "drive letters" (like F>, K>, M>), users can map drive letters to access Teams content. More details on file access options in Microsoft Teams provided in Chapter 4, "Utilizing Teams for Content Sharing and Collaboration."

Figure 1-2: Library of Files in Microsoft Teams

- Persistent Chat - Teams allows users to post questions, comments, and information into a chat window that is retained in the Teams channel specific to the conversation that can be accessed by others. Teams members can review current and past conversations at any time and see threaded conversations of members of teams and channels as shown in Figure 1-3.

Figure 1-3: Persistent Chat in Microsoft Teams

- Teams Calendars - Teams has group calendars where tasks, deadlines, milestones, and other details are posted for the group.

The group calendars are viewable side-by-side in user's Microsoft Outlook calendars so users can see personal calendars as well as group calendars from a single view. A combined calendar view of personal and team calendar events is shown in Figure 1-4.

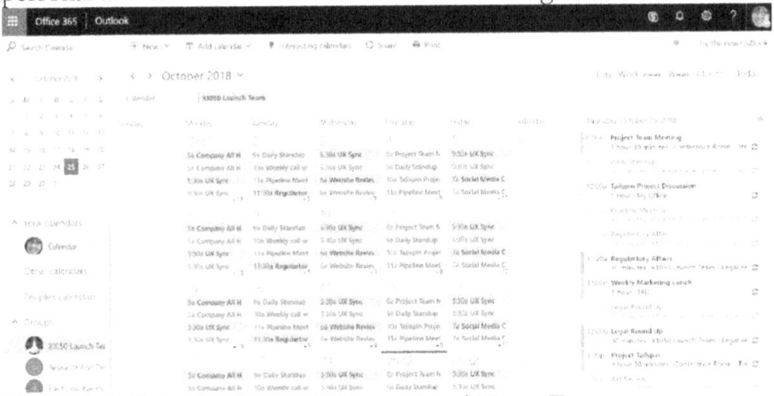

Figure 1-4: Calendar View of Personal and Team Events

- Video Integration - Office 365 includes Video upload and playback that allows organizations to retain full control over video content (ie: video captures of meetings, presentations, trainings, etc) that are controlled like any other Office 365 content including logon access and permissions. It's effectively YouTube videos, but with enterprise level controls and management. Videos can be linked in the Conversation View in Teams (as shown in Figure 1-5), stored as viewable files in a document library, or entire Microsoft Streams libraries can be linked off a tab in a team or channel. More details on this is covered in Chapter 7, "Getting Familiar with Teams and Leveraging Every Day Best Practices."

Figure 1-5: Microsoft Stream Video Streaming within Teams

- Integrated Project Planning - Office 365 enterprise licenses includes Microsoft Planner (shown in Figure 1-6) which is a project planning app similar to many 3rd party project and group planning apps, however Planner is built in to Office 365 and directly integrates with group calendars in Teams. Microsoft Planner helps teams consolidate milestones, due dates, and other project-focused deadlines within a user's familiar email, calendar, and content communications mechanisms.

Figure 1-6: Microsoft Planner Task and Group Assignment Tool

- Web/Video/Audio Conferencing - Teams integrates common Web Conferencing (as shown in Figure 1-7) along with video and audio conferencing just like Skype (and WebEx, Zoom, GoToMeeting, etc) has done for years, and brings collaboration and communications together into the single Teams interface.

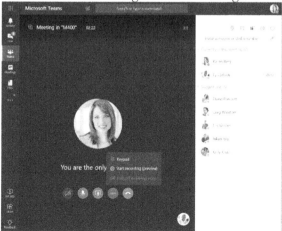

Figure 1-7: Web Conferencing with Microsoft Teams

13

- Telephony – One of the more recent integrations into Microsoft Teams that Microsoft has fully completed is the integration of Telephony that includes common phone system features like Auto Attendant, E9211, Transfer to Groups, voicemail, and the like. Microsoft Teams provides the ability for users to sit in Teams and make and receive regular PSTN phone calls right from within the Teams interface as shown in Figure 1-8.

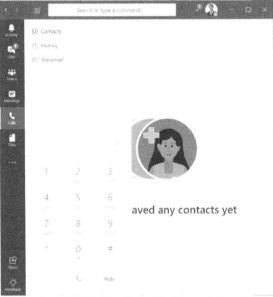

Figure 1-8: PBX Telephony Integrated into Microsoft Teams

Structured vs Unstructured Collaboration

One of the first things users who have worked with team collaboration tools in the past like Slack, HipChat, or the like realize is communications and collaboration is more <u>unstructured</u> than what organizations have traditionally been used to. When you look at the historical use of things like Slack, rarely is anyone telling anyone else how to format, structure, and manage the private and public channels, or set any rules and protocol in the usage. Users simply jump in, create a channel, invite users, and start chats and store information. Users "figure out" how each channel is structured and users participate in the communications adhoc. This adhoc method of site creation means organizations have hundreds if not thousands of channels or sites spun up, 60-80% of them are no longer used, and while some content is referred to weeks/months later, most of the sites and channels are never touched again.

14

Whereas organizations that have been utilizing SharePoint sites, Intranet sites, and formal document management tools in the past have built very structured templates, and implemented very clear protocols and processes for communications.. Each site these organizations have created follow specific standards and are (theoretically) used for months and years with long histories of content retention.

Organizations that jump in to use Microsoft Teams come in with completely different perceptions of what they will use Teams for and how Teams will be structured (or not). This is perfectly normal, and in fact in the best run Microsoft Teams environments, there's a mix of BOTH structured and unstructured content and communications.

Chats by definition are conversations that are adhoc and unstructured, whereas the portions of Microsoft Teams that leverage SharePoint pages and document libraries will tend to have more structure to them. That's actually been one of the biggest advantages Microsoft Teams has had for organization, in bringing together both structured and unstructured communications into a single tool and blending the benefits of both in improving modern intelligent communications within the enterprise.

When one looks at Microsoft's roadmap for Microsoft Teams, Microsoft is adding in more self-service features to Teams to support those functions that are less structured at the same time of adding more formal templates and workflow processes to Teams to support communications processes that need more structure.

Some think aloud that they will wait until Teams has new functionality added, however in reality, organizations are realizing that there is a new world of communications and collaboration that blends a bit of unstructured to allow groups of users to quickly self-service provision collaboration and communications spaces for short bursts of project engagements that have a bit of structure in them to allow enterprises to address regulatory compliance and data leakage protection.

With this blend and the anticipated ongoing changes in collaboration and communications, new practices and processes will frequently pop up and gain adoption. Organizations are consolidating the number of tools they own and use, while also working to improve the overall security and risk exposure the organization has in managing its content. All these changes enable organizations to leverage a broad functional and business integrated tool like Microsoft Teams in its current rendition, and finetune business processes as the tool is updated and enhanced by Microsoft over time.

Two Apps for All User Communications and Collaboration

With all of the functionality built-in to Microsoft Teams along with users already using Microsoft Outlook for emails, calendars, and contacts, these two applications take care of the intelligent communications and collaboration needs users typically have in an enterprise. And with these apps having full functionality no matter what platform the user is running (Windows, Mac, iOS, Android, or Web), organizations greatly simplify basic and sophisticated collaborative communications throughout the enterprise.

How About Microsoft Yammer

Microsoft continues to support Yammer and has repeatedly confirmed its commitment to keeping Yammer around for a while as many enterprises have standardized on Yammer to run their internal and external communications. For organizations that have not strategically adopted Yammer yet, what was unique about Yammer a decade ago isn't really unique for communications and collaboration tools anymore. With Microsoft Teams now effectively doing everything Yammer is best known for (and more), many organizations are simply standardizing on Teams for ALL of their internal and external communications and collaboration needs.

The reason an organization may choose to adopt and embrace Yammer:

- That key business partners it works with regularly use Yammer - many organizations regularly communicate with other enterprises that are heavily invested in Yammer, as such, communications with that other enterprise may likely dictate the use of Yammer for the organization.

- Yammer used to be better than Teams because Yammer provided virtually open collaboration and sharing of content and ideas, whereas SharePoint and (previously) Teams required you to be licensed, buy a software program, or do something special to collaborate externally. However by mid-2018, Microsoft released a free license offering for Microsoft Teams, now organizations can collaborate with anyone via Teams without being an Office 365 customer

Microsoft Teams Now and Into the Future

Microsoft has most certainly ramped up the development and feature release of Microsoft Teams in just a couple years, and provides an extremely robust tool that does more than just provide content collaboration and chat, but shared calendaring, project planning, telephony, and more.

Organizations that are already using Office 365 for their email should most certainly look to shift the dozens of one-off apps and solutions into a more consolidated platform under Microsoft Teams that provides tight integration, extensive security and compliance management, and is included in standard enterprise licensing and support.

Revising and Finetuning the Message

At Microsoft's 2017 Ignite Conference, Microsoft shocked the marketplace by stating that "Skype for Business is going away" and that Microsoft Teams will replace Skype. At the time, Teams was still a very new product, actually didn't do a fraction of what Skype did for organizations in terms of chat and most certainly telephony, so how could Teams replace Skype?

In the months following Ignite 2017, Microsoft backpedaled and said that Skype wasn't going to go away for a while, that organizations would be able to continue to roll-out Skype for Business in the cloud, leverage the telephony capabilities of Skype, and run both Teams and Skype to fulfill enterprise communications needs.

Where Teams Stands Today Relative to Skype for Business

Roll forward to today, Microsoft Teams has caught up in features and functions to Skype for Business, and Microsoft has a better message to provide the marketplace regarding Teams and Skype. Few things on the messaging:

- Microsoft Teams has matured significantly in feature capabilities and now does (pretty much) everything that Skype has done in terms of chat and telephony
- Microsoft rolled out a Free version of Microsoft Teams that now allows enterprises not licensed with Office 365 to participate in Teams conversations, meetings, and groups
- Microsoft's Teams Roadmap has more things "done/included" than in the "future release" column, that clearly makes it more full featured for organizations to shift to Teams as the primary/sole collaboration and communications platform.

The Simplicity of Teams

Microsoft Teams doesn't need to be complicated nor require a whole lot of user training. Unlike other tools of the past that have been completely disconnected from other business tools like email, calendars, file sharing, collaboration, or the like, Teams actually tightly integrates within Office 365's Outlook Web interface and any Office 365 security that an organization has in place. By tightly integrating with multi-factor authentication, legal hold, eDiscovery, Microsoft Information Protection (MIP) encryption, etc, it makes it a lot easier for users to get up to speed with Teams and for organizations to gain user adoption of the Teams platform in the enterprise.

2 ACQUIRING AND ENABLING MICROSOFT TEAMS

Most enterprises with Microsoft Office 365 licenses own the right to use Microsoft Teams, however there are a few steps to enable Teams for users (or selectively disable Teams for some users in the organization) along with setting up the initial structure of Teams for the organization.

Acquiring Microsoft Teams – Licensing and Subscriptions

For organizations that already own Microsoft's Office 365 Business or Enterprise licenses, they already own the license for Microsoft Teams. The Office 365 subscription plans that include Teams include:

- Business Essentials
- Business Premium
- Enterprise E1, E3, E5

However even organizations that don't own Office 365, Microsoft has a freemium version that supports up to 300 users with a major subset of the Teams features included. Some of the features excluded from freemium teams include:

- Meeting scheduling and calendaring (that integrate with Exchange and Outlook Online)
- Meeting recordings (that is part of Microsoft Streams)
- Audio conferencing (that is part of the Web/Video/Audio conferencing)
- Admin tools for managing users and apps
- Usage reporting

Effectively the freemium version lacks the integration that comes with the Office 365 suite, however as a free standalone version, users can conduct chat conversations, store and share content, and participate in group versions with users both internal and external to the organization.

For more details on the freemium offering, see: https://products.office.com/en-us/microsoft-teams/free

Enabling Teams for Users

For a user to use Microsoft Teams, they need to have a license enabled in the Office 365 Admin Console. For Office 365 tenants with Business and Enterprise licenses, Teams is enabled by default when the user is created. For academic and education tenants, Teams is disabled by default and needs to be enabled by the tenant administrator.

To enable or disable access, the tenant administrator can do so from the Office 365 Admin Console or from a PowerShell command in the following manner:

From the Office 365 Admin Console:

1) Go to https://portal.office.com and logon with administrator credentials.
2) On the left hand navigation bar, scroll down to Users and select Active Users.
3) Select the user (or users) you want to enable (or disable) Microsoft Teams by clicking the checkbox for each user.
4) Under "Product licenses" click Edit.
5) Slide the option for Microsoft Teams either On or Off to enable or disable Teams as shown in Figure 2-1.

Figure 2-1: Enabling Microsoft Teams in the Office 365 Admin Console

6) Click OK to set the setting.

For administration through PowerShell, do the following:
1) Open PowerShell as an Administrator.
2) Type `Install-Module -Name AzureAD`
3) Then type `Connect-AzureAD`
4) Make sure you have the "Microsoft Online Services Sign-in Assistant" installed https://www.microsoft.com/en-us/download/details.aspx?id=41950
5) Enter `Install-Module MSOnline`
6) `Connect-MsolService`
7) `Get-MsolAccountSku`
8) `$acctSKU="<plan name>`
9) `$x = New-MsolLicenseOptions -AccountSkuId $acctSKU -DisabledPlans "TEAMS1"` (where TEAMS1 would be the name of the plan in your tenant)
10) `Set-MsolUserLicense -LicenseOptions $x`

For more clarifying details on these sequences, see https://docs.microsoft.com/en-us/microsoftteams/user-access

21

Organization may choose to disable Teams for all users and selectively enable Teams for a small subset of users. This is commonly done during a pilot implementation and will help organizations manage the rollout of Teams. However since Teams is enabled by default, typically by the time the organization is thinking about a structured rollout, users are already using Teams in an adhoc manner, so the enabling or disabling of Teams may or may not be an effective in your organization.

Accessing Microsoft Teams for the First Time

There are two ways to access Microsoft Teams, one is through the Web interface and the other is by installing the Teams client and logging in from the client.

To access Teams from a Web URL, do the following:
1) Go to https://teams.microsoft.com
2) Type in your Office 365 logon credentials

To access Teams from the Teams app:
1) Download the Microsoft Teams app by clicking on "Download App" from https://products.office.com/en-us/microsoft-teams/group-chat-software
2) Choose the platform you are using (Windows, Mac, Mobile)
3) Once the Teams app has been installed, launch the application
4) Type in your Office 365 logon credentials

Web vs App Usage of Microsoft Teams

Microsoft Teams runs both from a Web browser and from an installed App. Unlike other client software where users may prefer to only run the application in either Web or App mode, interestingly with Microsoft Teams, many users switch between the two modes in the course of the day. When the user launches Teams (Web or App), the view looks the same, similar to what is shown in figure 2-2.

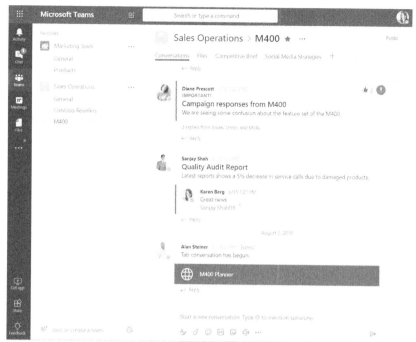

Figure 2-2: Common View of Microsoft Teams

The reason many users work from a Web browser is the Web mode provides a more seamless experience for users that have been working with Slack, Google, or other Web-only formats. Users who are familiar with a Web experience find moving between Web-based Teams to Web-based Calendars to Web-based Emails visually and functionally easy to use.

However some users that prefer an "app" like working out of the Microsoft Outlook client and the Microsoft Teams app will simply toggle and switch between the two apps. There hasn't been one single best practice evolving from usage, it has been one of preference and as noted, even some users using both interfaces interchangeably.

3 BEST PRACTICES IN SETTING UP AND ORGANIZING TEAMS AND CHANNELS

Once Microsoft Teams is licensed and enabled, the organization is ready to start creating its first Team environment and users can start getting familiar with the basic chat functionality of Teams. This may seem simple to just open up Teams and "let users at it," however it is helpful to put some thought into how the organization will organize and structure the Teams environment for the enterprise.

Many organizations find Teams is already in use and need to go back and retroactively put structure into the Teams environment, however it does not require wiping everything away from scratch and starting anew. Most organizations take the approach of implementing a "better" organized structure from a certain point forward so that old (unorganized) and new (organized) structures are used in parallel. Within a few weeks the new structure will have taken root and within a few months the old structure will be far behind with little or no lasting remnants.

Understanding Teams vs Channels

The first major concept in understanding Microsoft Teams is the idea of a team versus a channel. These are the two major organizing factors in

Microsoft Teams and when created, as shown in Figure 3-1, the team is the larger grouping (ie: X1050 Launch Team) and the channels are the subsets under the team (ie: Design, Digital Assets Web, Go to Market Plan, etc).

Figure 3-1: Sample Teams and Channel Structure

A Team is the top of the hierarchy for a group of users. This is where you assign user rights and access to the Team. Many organizations create a team for each department (ie: Marketing, Finance, Sales, Engineering), assign rights by the department users to that Team, and everything else is created as channels.

Some organizations choose to create Teams around projects or initiatives so that the structure is based around teams such as Finance-Quarterly Reporting, Marketing-Q4 Campaign, or Northwest-Sales.

The key factor in choosing the root Team structure is the fact that the subset under Teams (called channels) does not allow you to add or remove users to access the channel. So if you create a channel like HR Records under the Finance Team, anyone in the Finance Team will have access to the HR Records channel.

This might not matter for an organization as everyone in Finance may also by default typically have access to HR Records, however there is currently no way to hide or give granular access to a channel that is different than the Team.

However the granularity of security access, or the ability to make channels private or public is easily addressed by simply having a mixture of Teams based on departments AND Teams based on topics. When a series of topics ("channels") all flow under a common group of users ("team"), then a team is created with a number of topical channels underneath it. However when a topic requires a unique set of security controls (ie: limited to just a handful of users), then the topic IS created as a team.

Here are some examples of teams and channel structures for organizations:

Team=Sales
Channels=Customer A, Customer B, Customer C...
In this case, the Sales Team members are all of the people in the organization in Sales (sales managers, sales representations, customer service assistances to sales) and each customer's information will be separated into individual channels. This provides a simple view for anyone in sales to have access to just ONE Team that they can find any and all customer accounts. And each customer account will have separate conversations, files (for quotes, proposals, contacts), and account planning records.

Team=Customer Z
Channel=(none)
In this example, a specific customer (in this example, Customer Z) has their own Team. While all other customers may be channels under the Sales team, this one customer is isolated as their own team. This might be done because this customer has sensitive information that only a handful of sales representations should have access to (not everyone in the Sales team). Or simply the size and scope of work the organization does for this customer dictates a need to raise the attention of a dedicated Team structure for this one customer. For simplicity reasons, organizations find it easier to put all customers under a common "sales" team, but this is one of those cases where structure can be overwritten with adhoc flexibility, one of the nice things about Microsoft Teams in providing variations to absolute structure.

Team=Marketing Projects
Channels=Q1 Campaign, End of Year Campaign, Digital Marketing Initiative, etc
In this example, everyone in the marketing department will be given rights and access to the Marketing Projects team. Each campaign will have its own channel for files, conversations, and details separated for each campaign.

Team=FY19 Yearend Report
Channel=(none)
In this example, the organization may have a separate team created for just 2-3 individuals that need to discuss the Yearend results that'll generate the financial reporting information to shareholders and investors. In this case, the Team will be limited to just a handful of users, not all users in finance, or all executives for that matter. All team conversations are focused to this yearend reporting, so separate channels are unneeded at this time. All documents and conversations are focused to this one topic isolated to a specific set of users. In this case, the team is the project and conversation.

As you've seen in these examples, when possible, try to create a limited number of Teams and use channels to create subsets within the Team to minimize the number of Teams a user is a member of. However create individual Teams typically when user access needs to be limited due to confidentiality or purely isolated access reasons.

One thing to consider is the portability of teams and channels. Teams and channels, once created, cannot be "moved" to nest under different teams or channels. For an organization that might have created a team called "Bob's Sales Accounts" with a list of 10 accounts, if in a subsequent year Bob has a handful of his accounts move to another sales representation, there's no easy way to move all of the conversations, documentation, and details from Bob's Team to someone else's Team.

This is why creating an all-encompassing Sales Team provides easier access to content if teams change between users. However don't get too caught up on creating the "perfect" team/channel structure as Microsoft Teams is intended to have flexibility for use by different organizations.

It has also been found that structures in Microsoft Teams change over time with organizations as users work within Teams, so start with a basic framework based on initial thinking and best practices, and then let users modify and change their usage and Teams structures as they see fit.

Creating a Team

To create a team, within the Microsoft Teams app or Web interface, at the bottom left where all of the Teams are listed, click on "Join or Create a Team" as shown in Figure 3-2.

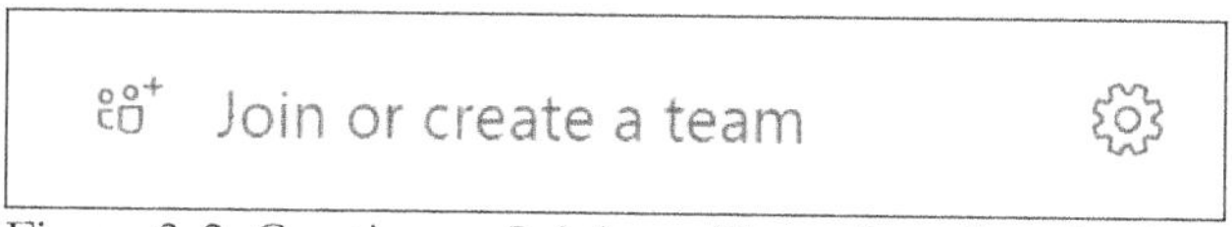

Figure 3-2: Creating or Joining a Team Interface Setting

Click to Create a Team and enter in the Team name, a Description of what this team is about, and choose to make the team private, public, or Org-wide (private meaning that only a team owner can add other members; public meaning that anyone can join and participate in the Team; or Org-Wide where everyone in the organization will automatically be added to this Team).

Click Next and enter members of the Team and build out initial content for the team.

Creating a Team from a Template or PowerShell

Microsoft Teams provides a mechanism where Teams can be created from a Template. This is helpful for organizations that want additional structure when Teams are created such as having specific Microsoft and 3rd party components added each time a team is created.

Teams as well as channels and other content can be created through PowerShell scripts, providing another mechanism for automation in the creation of Teams and content. For more details on PowerShell commands available, see: https://docs.microsoft.com/en-us/microsoftteams/teams-powershell-overview

Creating a Channel

Once a team is created, team members can create channels under the team. This is optional as some organizations will find the team IS the sole and primary place for files and conversations to occur for that team. However if the team will be broken down into smaller subsets (such as Customer A, Customer B, Customer C under the Sales Team) then channels should be created.

To create a channel, click on the ellipse (...) to the right of the team name, and choose "Add Channel". You will be prompted to give the channel a name and optionally give it a description. Within the channel, you can now have conversations that remain within this channel, and upload files and content to the channel library that are files specific to this channel.

Using Chat for Persistent Communications

With a team created (and potentially channels created), users can now carry on chat conversations in the team or channel. These are conversations that are viewable by all members of the team. As an example, within the Marketing team, a conversation might be about an upcoming digital marketing campaign that will go live in the next couple weeks where members of the conversation want to ask questions and check in on the status of the campaign.

The conversations are persistent, so they are treaded and remain in the team or channel for others to view down the line.

A user can "Start a new conversation' simply by typing in content under the "Conversations" tab in the area shown in Figure 3-3.

Figure 3-3: Starting a New Conversation

What users will find is when they type content, if they press the <Return> key, it will post the conversation, so by default, the content typed is a single paragraph. However, by simply clicking on the "A" (with a paintbrush) icon on the bottom toolbar (shown in Figure 3-4), it brings up a rich text entry form where the user can now type multiple paragraphs, add bold/underlining, change colors, add in bullets, link files or other content, and change general formatting. This rich text format entry provides a lot more flexibility in information entry over the simple few words / single paragraph of the default entry.

Figure 3-4: Toolbar to Choose to Type in Rich Text and Multiple Paragraphs

Also within the conversations thread, users can add in emoji, stickers, or link content.

When entering in conversations, a user can call attention to other team members by doing @username (like @rand or @chris) which is called a "mention" so that when the user logged into Teams, they can go straight to the conversations where they are specifically mentioned and called out.

4 UTILIZING TEAMS FOR CONTENT SHARING AND COLLABORATION

Once teams and channels are created, the team of users typically find a key component of communications is sharing files and content. Microsoft Teams provides an extensive mechanism for saving, sharing, co-editing, and managing files of all types. For those familiar with Microsoft SharePoint, SharePoint libraries, and content sharing in SharePoint and liked it, Teams leverages SharePoint Online for content management and collaboration.

For those who used to use SharePoint years ago, hated it, and SharePoint might be a bad word used around the enterprise, assuming it was months and maybe even years ago that SharePoint was formerly used, do know that SharePoint these days is a much more streamlined technology and offering. First of all, Teams if not based on SharePoint whatsoever, Teams is a completely newly developed cloud-first environment. The only part of SharePoint brought forward to Teams is the file and content components, and now that SharePoint Online is in the cloud, Microsoft has 100% compatibility and support for Macs, mobile, and Web platforms beyond Windows. So it is a completely different and greatly improved environment.

Accessing Files within the Teams Interface

Within teams and in channels in Microsoft Teams is a tab called "files" where users can upload files for shared team and channel user access. Uploaded files could be Word documents, PDFs, audio or video files, or other content that users want to share and collaborate on with other team members.

Files can be downloaded, opened, linked and shared with others, effectively making the file(s) available to others both internal as well as external to the team dependent on security permissions. For external access, more details are covered in Chapter 9 "Managing and Administering Teams" in the section on "Enabling External Access to Teams Content."

By clicking on the "…" ellipse next to any file, users can open the file or folder in a Microsoft Teams view or in a SharePoint view.

The Teams view displays the file or folder within the standard Teams interface, whereas in the SharePoint view, a new browser page is opened providing a view of the content from within a SharePoint Web view. Most day to day access to content will be within the Teams view as shown in Figure 4-1 as the user will open and access content within Teams.

Figure 4-1: View of Files in the Default Microsoft Teams View

However when sharing a link to content, or for accessing files outside of Teams, then the SharePoint view effectively provides a more generic Web-page view of content that is more universally accessible from any browser. The Web-page view makes it easier to share links with users who may not have the Teams client installed on their system and/or work outside of the Teams environment.

Accessing Files within the Outlook Interface

Files that have been uploaded to Microsoft Teams can also be accessed within the Microsoft Outlook interface in the Groups section for each Team. When navigating to groups and selecting a group or team, there is a "Files" tab at the top of the page. Clicking on any of the files in the file library will allow a user to upload, download, view, and preview files.

What the user will notice is if they are in the Microsoft Outlook client software, by clicking on "files", a Web page is launched exposing the Teams file folder. However if the user is using the Microsoft Outlook Web interface, effectively accessing Outlook from a browser, then when the user clicks on Groups / Files, the files page will just show up in the Web browser that the Outlook web view was being access.

This seamless shifting from Outlook (web view) to Files (web view) to other web views is one of the nice benefits of working from a Web experience as opposed to working from within "apps." Users who have worked heavily with Google Docs, Google Drive, Box.com, HipChat, Slack, and other web-focused applications find the uniformity of using the Web browser for content access more seamless.

This becomes one of those preference things where some users will gravitate toward using apps (that opens up multiple applications and interfaces that users "toggle" between) versus users who use the Web interface and navigate within a browser to access various bits of content.

By using Microsoft Outlook / Groups, or using Microsoft Outlook Web and Groups, instead of forcing users to toggle over to the Teams app, or switching over to a SharePoint page, or launching yet another 3rd party file / collaboration / communications tool, users can simply STAY in Outlook (where their emails, calendars, and contacts are) and click on Groups and have access to Teams/Files.

Viewing Files in List View vs Tiles View

Another variation in user views is the ability to access the file directory to show files as Tiles or in a List. Users that are more used to a common Microsoft Windows or Mac interface with a list of files may find the List View more familiar to them. Users that use mobile devices and tablets that have gotten used to seeing files as icons may find the Tiles view to their preference. A user can switch between List and Tiles views by simply clicking the List/Tiles in the upper right of the page that switches views of the content. A comparison of the view of the same file folder, but with a different Tile versus List view is shown in Figure 4-2.

Figure 4-2: Tiles and List View of Files in Microsoft Teams

Mapping SharePoint Online and Microsoft Teams Folders to File Explorer Drive Letters

A common request organizations using Microsoft Office 365 have is being able to MAP their SharePoint Online and Microsoft Team folders to drive letters (F>, G>, K>) like users have used on-premise for years.

With the ability to map drive letters, organizations that are migrating off on-premise fileservers and fileshares to Office 365 can simply drag/drop content from on-premise storage systems to Office 365, and (very commonly) maintain the exact same directory structure as the users are familiar with. K-Drives remain K-Drives, M-Drives remain M-Drives, P-Drives remain P-Drives, etc.

Organizations that may have tried this process in the past may have had a difficult time getting mapped drives to Office 365 to work. A while back, the feature just didn't work. Some time back, Microsoft posted a guide on how to bypass some of the security features to allow the functionality to work, however any organization serious about security (which should be EVERY organization) wouldn't allow "saved credentials" and will likely have multi-factor authentication enabled that thwarted the ability to make folder remapping work. However with recent workarounds that we've been able to document, there is a better method that allows for drive mapping to Office 365 file stores in a secured manner, even when Office 365 multifactor authentication is enabled.

The step by step process of mapping drive letters to SharePoint Online (SPO) shares is as follows:
1) Logon to a Windows 7 (or more recent) computer like normal
2) Launch Microsoft Word 2016 (or more recent)
3) Do a File / Open / Sites / and click on your organization's SharePoint Online (SPO) site.

If you don't have SharePoint Online connected to Word yet, click on the "+ Add a Place", select SharePoint Office 365, logon to Office 365 using your Office 365 credentials, and it'll add your organization's SPO and OneDrive to Word.

If you have SharePoint Online connected to Word, or now have SharePoint Online connected to Word, you will now see OneDrive and SharePoint Online connected to Word similar to what is shown in Figure 4-3

Figure 4-3: View of Microsoft Word Connected to SharePoint Online and OneDrive

4) After you click the Sites – (Your Company SharePoint Online site), paste in the URL for the SharePoint library you want to map a drive letter to (up to the Shared Documents\ piece, cutting off the forms\allfiles.aspx

As an example, the full URL to the Documented folder might be: https://companyabc.sharepoint.com/sites/KnowledgeSharingSite/Shared%20Documents/Forms/AllItems.aspx as shown in Figure 4-4.

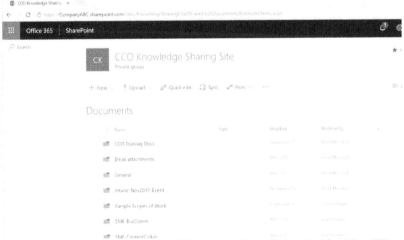

Figure 4-4: Copying the Web URL of the SharePoint Folder You Want to Map

Cut/Paste the following:
https://companyabc.sharepoint.com/sites/KnowledgeSharingSite/Shar
ed%20Documents/ as shown in Figure 4-5

Figure 4-5: Pasting a portion of the Web URL of the SharePoint Folder
You Want to Map

5) Then click Open (that'll open the SharePoint Online or Microsoft Teams folder (and complete the authentication piece). Users could just work from this point by doing a File/Open, File/Save files right from Word (Outlook, Excel, etc) right to the Office 365 folder.

6) But to Map a Drive Letter, go one extra step. At the bottom of the dialog box to the left of the Open button, click on Tools and select "Map Network Drive" and paste in the URL of the file share again as shown in Figure 4-6.

Figure 4-6: Mapping the SharePoint Folder to a Mapped Drive Letter

As an example, select Z> for the drive letter, also select "Reconnect at Sign-in" (although it doesn't automatically reconnect Z> after reboot (more on that later)), and click Finish

7) Now in File Explorer, you will have a Z> mapped, similar to what is shown in Figure 4-7 that looks like every other fileshare that you can drag/drop files, create folders, copy folders, etc...

Figure 4-7: Mapped Driver Letter in Windows Explorer View

Note: You cannot directly map a drive letter in File Explorer without first going through this Microsoft Word file/open process ONCE and setting up the drive letter, so this is a needed process to get it to work. The reason you can't directly Map a drive letter in Windows Explorer without first going through this Microsoft Word process is you haven't authenticated to Office 365 applications yet, so the attempt to map directly from File Explorer errors with a "The folder you entered does not appear to be valid. Please choose another." and it will not directly map the drive share.

With a SharePoint Online folder mapped to a drive letter, you can now open another File Explorer window of your existing F>, K>, P> drives and drag/drop content from one folder to another!

Remapping the Drive Shares after a Reboot

After mapping a drive letter to a SharePoint Online share, what you will find is when you reboot your system and open File Explorer, the drive letter you mapped cannot be accessed. The icon shows the drive share is disconnected.

The way to get your drive letter remapped again:
1) Launch Microsoft Word
2) Do a File / Open other Documents / Browse as shown in Figure 4-8.

Figure 4-8: Remapping a Drive Letter After Reboot

3) At the Open dialog box (that defaults to your C>), click on the drive share(s) that you had mapped as shown in Figure 4-9 and it will automatically re-authenticate your session in the background and immediately reconnect your drive letter(s) again for full access and no extra steps!

Figure 4-9: Automatic Reauthentication for Mapped Drives

So a one time series of steps to get the drive letter(s) re-mapped (through Word) and then thereafter it is a fully accessible File Explorer share until you reboot your system again.

Editing and Co-Editing Content

With content stored in Microsoft Teams, users can open and edit Office documents either in a Web session (that Microsoft calls their "Online" version of Word or Excel) or the user can launch the full application (ie: Word, Excel, PowerPoint) and edit the content within the application.

When editing a file in a Web browser, Microsoft supports browser types on various systems like Internet Explorer, Safari, Firefox, and Chrome on Windows, Apple Macs, and mobile phones and tablets. And from an application perspective, the files can be downloaded and edited in Office versions on Windows, Macs, and mobile devices as well.

Co-Editing is when more than 1 person is editing the same document at the same time. Office 365 provides the ability of co-editing so that multiple users can be modifying and editing the same document at the same time. Simultaneous editing accelerates the time it takes for content to be reviewed and modified. Instead of writing and editing content, then saving it and sending it to someone else to review and edit, and sequentially sending content back and forth, co-editing provides simultaneous content access and modification.

When multiple users are co-editing a document, the changes of all individuals are saved, and users can see edits being made by other users by seeing the user's name show up on the document to denote where in a document the user is accessing and editing similar to what is shown in Figure 4-10. Content is automatically saved so that content edit tracking and identify modifications to documents.

Figure 4-10: Simultaneously Editing Documents in Office 365

Content Management and Version Controls

As users are sharing and co-editing content, Office 365 provides version controls so that users can view changed content of each and all users, and with appropriate security permissions can accept all edits and changes, eliminate any and all changes made by others, or simply view the edits and choose to modify the edits that were made by others.

Version history provides users the ability to pick the most current version of the document to modify and edit, or review older revisions of the document to compare historical edits as well as accept or decline edits made of other users. Version history appears in Office 365 similar to what is shown in Figure 4-11.

Figure 4-11: Version History View in Office 365

The version control settings as well as changes that are tracked and displayed provides users the ability to manage modifications to content.

Sharing Content Internally and Externally

Because Microsoft Teams has the ability for administrators to enable access by non-employees (ie: external users), content can be shared both with internal employees and team members as well as with people outside of the organization.

For external access, more details are covered in Chapter 9 "Managing and Administering Teams" in the section on "Enabling External Access to Teams Content."

Enabling Security Protections to Files

Microsoft Teams and content stored in SharePoint Online folders have standard permission settings that can be set for users to have full access, read only access, or no access as normal content permission settings.

For organizations that want to provide more sophisticated data leakage protection with more than 15 detailed permission settings (like print / no print, copy-paste / no copy-paste, expiration dates on files, and the like), organizations can get a content security license that provides what Microsoft calls "Microsoft Information Protection," or MIP.

More details on data leakage, information protection, and content retention is covered in a book titled "Handling Electronically Stored Information in the Era of the Cloud" that you can freely download at http://www.cco.com/our-publications.htm

5 EXTENDING TEAMS FOR WEB, VIDEO, AND VOICE CONFERENCING

Microsoft Teams replaces Microsoft's Skype for Business functionality for Web, Video, and Audio conferencing. While Microsoft has made statements that they will continue to support Skype on-premise and existing Skype Online environments, Microsoft is no longer onboarding new Skype for Business Online environments for small enterprises with fewer than 500-users, and the writing is on the wall that Microsoft Teams is Microsoft's platform for future Web and online communications.

For early adopters of Microsoft Teams who found limitations in what Teams could do in terms of Web sharing, telephony, and the like, after several rolling updates, Microsoft has quickly brought Microsoft Teams up to parity with the capabilities organizations relied on for Skype for Business Online.

With Microsoft Teams being Microsoft's platform for Web, Video, and Audio conferencing, this chapter covers the functionality available in Teams for conferencing functionality.

Joining a Web Conference in Microsoft Teams

To join a scheduled Web Conference in Microsoft Teams, a user simply clicks the "Join Microsoft Teams Meeting" as shown in Figure 5-1 in the meeting/calendar invite just like joining a Skype, WebEx, Zoom, or other web meeting solutions of the past.

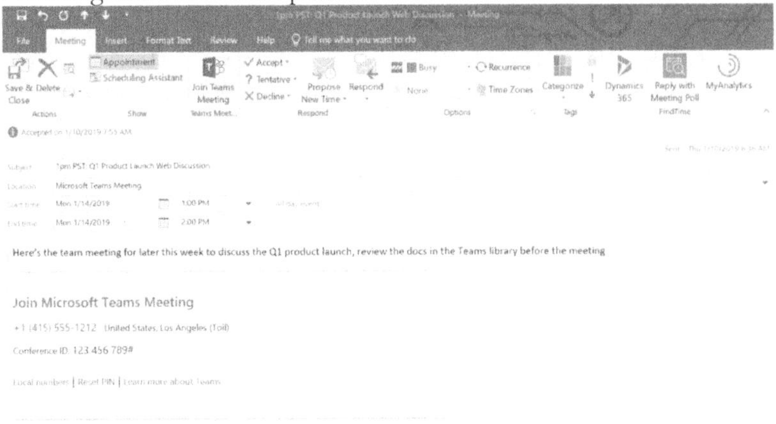

Figure 5-1: Joining a Meeting from a Calendar Invite

If the Microsoft Teams app (Windows, Mac, iOS, Android) is installed on the system, the Web meeting will launch within the Teams app. If the Microsoft Teams app is not installed on the system, the user can launch the Web meeting from within a browser. Functionality wise, it works the same regardless of running a meeting using the Teams app or from a Web browser.

Once in a Web meeting, the attendee has the ability to mute or unmute their microphone by clicking on the microphone icon, upload content that will be shared with others, turn on or off a camera to participate in the Web meeting as a video participant, record the meeting, or hang-up as options on the control toolbar shown in Figure 5-2.

Figure 5-2: Microsoft Teams Control Toolbar

Recording Teams Meetings

One of the options in the Teams meeting control bar is to record the Teams meeting which is helpful so that if someone is not able to join the meeting, they can watch a video of the session that includes the audio

44

conversations as well as see any materials presented during the meeting.

To record a meeting, once the meeting has started, the user can click the "…" ellipse on the control bar and select "Start Recording" as shown in Figure 5-3.

Figure 5.3: Starting a Recording in Teams

When the session is over, the user can click on "Stop Recording" that will stop the recording and process the recording and automatically upload the content in Microsoft Streams. A link to the video of the session will automatically be posted within the "Conversations" thread of Teams similar to what is shown in Figure 5-4. Users of the team can click on the recording and watch the content. Teams does a voice to text transcription of the meeting, so within a few minutes users will not only have an opportunity to watch/listen to the meeting, but can search words from the meeting and jump straight to the part of the meeting where specific words were mentioned. This helps participants navigate through a long meeting to find key words like "401k" or "benefits" or "Project X" or the like.

Figure 5.4: Link to Recorded Meeting Posted in Conversations

Adding New Participants to an Existing Teams Meeting

At any time during a team meeting, others can be added to the session. If the participant is a member of the team or channel, when the user clicks on the upper right people looking icon (the Participants Panel), it will open up a list of members of the team/channel that can be added.

If you want to add someone external to the team, when you are in the Participants Panel list from the upper right, you can further click on the chain links looking icon that'll "copy join info" and you can then email that information to someone who is not a member of the team or channel. That person can follow the "Join Microsoft Teams meeting". Upon clicking to join the meeting, the user will get a notice that "someone in the meeting should let you in soon…".

Participants in the Team meeting will get a notification in the team meeting that someone is "Waiting in the lobby" as shown in Figure 5-5. They can press X to deny the person access, or they can click the checkmark to allow the person into the meeting.

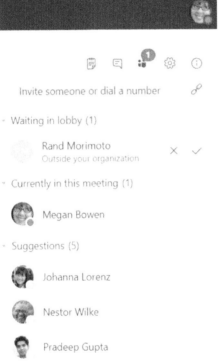

Figure 5-5: Person Waiting in the Meeting Lobby

Muting and Removing Participants

Within the Microsoft Teams meeting interface, participants can mute and remove other participants by opening up the Participants Panel, clicking on the participants name, and click the "…" ellipse. You will have the option to "mute participant" or "remove participant".

Creating Meeting Notes for a Meeting

Teams also has a "Meeting Notes" function that is helpful for notes to be created, action items noted, key points or meeting minutes created during the meeting. To create meeting notes while in the meeting in the upper right of the Teams meeting will be a paper looking icon that is Meeting Notes. Click on the Meeting Notes icon and click on "Take Notes".

The meeting notes interface provides the ability to take notes, create sections to highlight other information (like action items), insert in pictures and links, highlight text, create bulleted lists, and utilize rich text for formatting similar to what is shown in Figure 5-6.

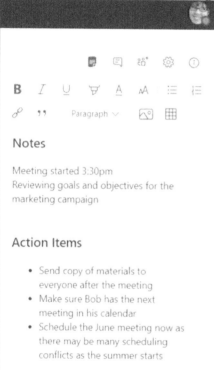

Notes

Meeting started 3:30pm
Reviewing goals and objectives for the marketing campaign

Action Items

- Send copy of materials to everyone after the meeting
- Make sure Bob has the next meeting in his calendar
- Schedule the June meeting now as there may be many scheduling conflicts as the summer starts

Figure 5-6: Taking Notes in Microsoft Teams

The meeting notes are automatically posted in the Conversation thread for the team/channel, thus making the notes available to all participates

Starting an Impromptu Web or Video Conference

Microsoft Teams has the ability of users starting a Web/Video/Audio meeting at any time using a "meet now" feature. From within the Microsoft Teams "conversation," click on the camera looking icon at the

47

bottom of the Conversation page, similar to what is shown in Figure 5-7, that will launch a Meet Now Web session.

Figure 5-7: Choosing to "Meet Now" to Start an Impromptu Meeting

Within the meeting, users can add others, share and present content, turn on cameras and make it a video session, record the session, etc just like any other Teams Web meeting.

Scheduling a Web or Video Conference

As with all conferencing platforms, the ability to schedule a meeting and invite participants in advance is an important function, and of course Microsoft Teams provides a way to schedule meetings. For users that have installed the Microsoft Teams app on their system (Windows or Mac), when the user is in Microsoft Outlook and scheduling a meeting, there will be a "New Teams Meeting" button on the ribbon at the top of Outlook, similar to what is shown in Figure 5-8, that the user can select to start creating a Web meeting.

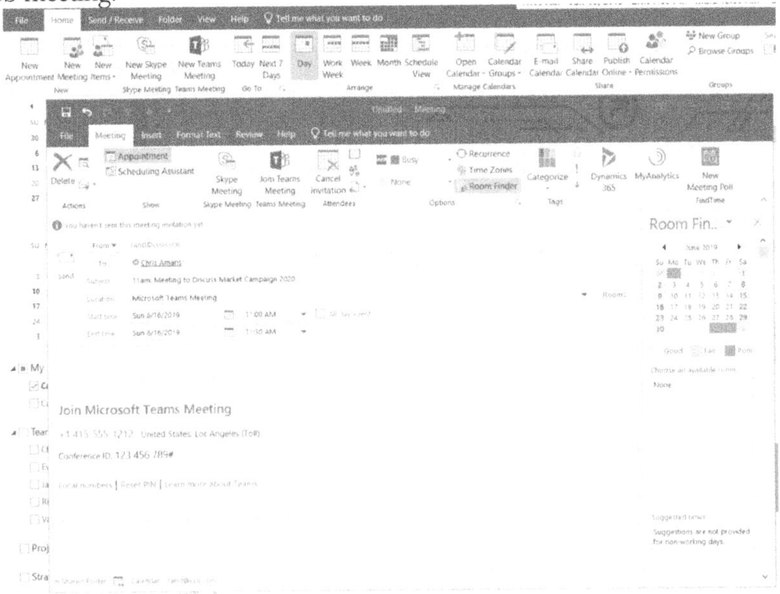

Figure 5-8: New Teams Meeting Button in the Outlook Ribbon

Alternately, if the user has already starting creating a meeting, there will be a "Teams Meeting" button at the top ribbon to add in meeting "join" information.

Note: Some users comment that the "New Teams Meeting" button does not consistently show up on their ribbon. This is commonly caused by the order in which the Outlook and the Teams clients are launched on the system. If the Microsoft Teams app is launched first and THEN the Outlook app is launched, then the "New Teams Meeting" button always appears on the ribbon. However if the Outlook app is launched first, and THEN the Teams app, the user might not see the "New Teams Meeting" button show up in the ribbon.

6 MAKING TEAMS YOUR CLOUD PBX PHONE SYSTEM

Beyond just Web, Video, and Audio Conferencing, Microsoft Teams can also be an organization's phone system. Just a few years ago, many organizations would have never considered moving their phone system to an outside service because of how business critical "phones" were to day to day commerce. However roll forward a few years and phone systems have given way to email and other digital methods of communications that phones are no longer the sole or even the primary method of communications.

Given that organizations have moved their email systems to Microsoft Office 365 for cloud-based email communications and have found Office 365 email to be reliable and secure, the follow-on shift by these organizations has been to move their phone systems to the Office 365 cloud as well.

Microsoft Team's Cloud PBX provide all of the normal business phone functionality such as auto-attendant, call routing, dial by name, voicemail, etc that organizations are used to. The feature and function parity is what has made organizations adopt the cloud being that they can "do the same thing in the cloud" as they've been doing on-prem for years.

Porting Existing Phone Numbers to Microsoft Teams

One of the first questions users have about using Microsoft Teams for their phone system is whether the users and the company can continue to use the same phone numbers they have used for years. In the case for organizations in the United States, phone number portability between fixed and mobile phones is something organizations and users are assured of since 1996 (and updated in 2003) mandated by the Federal Communications Commission (FCC). It is commonplace for organizations to port over all of their business and user phone numbers from their existing phone carrier to Microsoft Teams in the United States.

Canada and South Africa also provide assurances to businesses and consumers to port their phone numbers between carriers. This portability of phone numbers assures businesses and users that the phone numbers they have had for years can be ported to the Office 365 cloud.

For organizations in other countries, porting of phone numbers can typically be done between mobile carrier providers, but not always assured for fixed land lines, however this is taken up country by country. It is becoming more common for phone carriers around the world to allow organizations the assurances of phone number portability.

However in cases where phone numbers are not assured portability, this has rarely been a showstopper in preventing organizations from shifting their phone systems from on-premise to Microsoft Teams. This is helped heavily by the fact that very few organizations depend on their phone systems as their primary method of communications, and for any organization that has been around for a long time, they have likely undergone phone number changes in the past.

When businesses move across town or have area codes in their locations updated, the organization has historically just announced phone number changes, and leave an automated call notice that the older numbers have changed.

Back when phones were the number one and sole method of communications, a phone number change was something that required a lot of advance communications, however in this day and age when organizations have Websites, email, and other mediums that customers can contact them, phone number changes have less of a business impact than before.

And of course, for organizations where phone number porting is assured by law, the transition between carriers can be done with seamless cutover.

Roaming with Office 365 Cloud PBX

What organizations find once they switchover to Office 365 for their phone system is the ease in which the organization can move their offices (without switching phone numbers) and how easy it is for employees to get phone calls from remote offices, home offices, or right off their laptops or mobile devices.

Since Office 365 telephony simply requires an App installed on a system (Windows or Mac computer, Tablet, or mobile phone), as long as the app is running, the user's "phone" is available to receive incoming calls or for the user to make outgoing calls.

For the field representative or executive who is always on the go, having their "phone" follow them just like receiving emails from anywhere provides a method of better connectivity and communications.

For organizations using Office 365 for their phone system, when the organization needs to move its office to a different location, as long as the new facility has Internet connectivity, ALL phones, conference room phones, receptionist phones, and employee phone numbers remain the same in the new location.

For organization that expand into new office buildings, for organization that acquire other businesses and fold those organizations into the parent company, NO phone calls need to be made to the "phone company" to add phone lines, phone circuits, rewire connections, or add phone services. The I.T. team simply adds an Office 365 license to the user, assigns the user a phone number from the organization's pool of phone numbers, and a new user is added to the phone system of the organization in seconds.

Licensing CloudPBX Functionality

Licensing Cloud telephony is frequently confusing for organizations because there are a couple different licenses that "sound" like all that the organization needs to get. However once you understand the components, it's not that difficult.

The base license all users of Office 365 need to get is a Microsoft Office 365 license. This comes in the form of an Office 365 Business, Business

Premium, Enterprise E3, Enterprise E5, or the like. This license provides the user access to Office 365 email, files, Web conferencing, security components, etc based on the license purchased.

The thing that is more challenging is that those users who will get a "phone" in the cloud ALSO need to have a "PSTN license" as well as a Dial Plan to be able to make outbound calls.

- The PSTN license allows users to do person to person (app to app) calling and host web conferences. While Office 365 licenses that include Microsoft Teams licenses for users to host Web conferences along with allowing video and audio conversations during the Web conference, technically for users to simply do voice to voice "calls" outside of a Web conference, the user needs a PSTN license. These PSTN licenses come in the Office 365 Enterprise E5 licenses or can be purchased separately just for the users that need to do person to person (app to app) calls.

- BUT for users that need to ALSO dial beyond person to person (app to app) such as to make an 800# phone call, call someone's mobile phone, or call a business or user that doesn't have Microsoft Teams, those users need to have a "DialPlan". The DialPlan are like the "minutes" users buy/get when they buy a mobile phone plan. The DialPlan might include 3000 free minutes or "unlimited calling" but effectively a Microsoft Teams user needs a DialPlan to make traditional phone calls.

- The DialPlan in Office 365 runs around US$12/user/month for effectively 3000 domestic "minutes" of outbound calls

- The International DialPlan runs US$24/user/month for 3000 domestic minutes PLUS 200 international minutes

Note: Actual cost and the # of minutes varies, make sure to validate the actual cost and offering based on current licensing services from Microsoft.

Beyond the individual minutes assigned to each user, organizations can commonly buy "pooled minutes" from Microsoft so that users that exceed their allotted domestic or international minutes during a month can overflow to using prepaid minutes from the company.

Do note, every country has a different DialPlan offering, check Microsoft's documentation at https://docs.microsoft.com/en-us/microsoftteams/country-and-region-availability-for-audio-conferencing-and-calling-plans/country-and-region-availability-for-audio-conferencing-and-calling-plans for details.

Rightsizing Phone Licensing and Costs

When organizations price out the cost of migrating from an on-premise phone environment to a cloud-based environment, they make the assumption that EVERY employee in the organization still needs a phone. When that is done, the cost of cloud phones might seem costly.

However in this day and age, since phones are not a primary method of communications, many roles in the organization have NO need for a phone. As an example, a manufacturing organization with 70% of the employees working on factory floors likely have never received a direct phone call into the organization in the past year, two years, or ever for that matter.

Retail sales clerks may answer incoming phone calls to a store, but highly unlikely each sales clerk needs their own phone and personal phone number for incoming and outgoing phone calls.

Even I.T. personnel may find that while they get phone calls and make phone calls (to vendors, to business partners, etc) that they rarely make the phone call from an office or desk. These days, users commonly make calls from their mobile phone, or incoming calls go to the I.T. person's mobile phone. Thus the need for a desk phone for inbound and outbound calls may not be necessary.

The best way for organizations to rightsize the number of actual Cloud PBX phones and dial plans needed is typically done by breaking the organization into three separate groups:

1. Employees that Must Have a Phone: This would be those users in the organization typically belonging to the accounting and finance departments, executives, and direct sales associates. These are individuals that specifically need a phone number for people to call them on, and actively receive phone calls on a regular basis.

2. Employees that Have No Need for a Phone: These are individuals in roles where the user never receives a direct phone call, such as factory workers, farm workers, retail sales associates, or the like.

3. Employees that May or May Not Need a Phone: The balance of employees might fit in a third group that may straddle possibly needing a phone or not needing a phone. The organization can assess usage and choose to assign or not assign phones and dial plans to these users.

One of the easiest ways to determine who fits into what category is to dump the call logs off the existing phone system for the past 3 months and run a report that notes who has made and received phone calls in the past 3 months that have lasted more than 30-seconds. The 30-second marker for the length of the call attempts to weed out "junk calls" that are answered and quickly hung up as opposed to "real" phone calls where someone will be on the call talking to someone for a length of time.

Do note, when tracking calls, make sure to monitor only external source and destination calls. Internal office to office and employee to employee calls are handled without the need for a dial plan, so employees within the organization can make phone calls to one other without the Cloud PBX license.

Also, a best practice is for those employees that fit in the middle that may or may not need a phone, best thing to do is NOT issue the individual a phone, see how long they go without requiring a phone to conduct business, and if they don't actively make or receive phone calls, the user can continue without a Cloud PBX license and dial plan.

Shared Phones in Microsoft Teams

For organizations that have groups of employees that may need to make outbound phone calls such as for emergency purposes on a factory floor, or even readily available phones in a break room for employees to make phone calls, the organization can implement shared phones throughout work spaces. A handful of shared phones is much cheaper than buying dozens or hundreds of licenses for users that never actively use the phones or have a need for direct phone numbers.

The shared phones will be available for users to dial-out and make phone calls. The shared phones may also be incoming lines for a retail store or facility where the call is destined to "someone" in the organization that can answer the call and answer questions, or put someone on hold and find the person that the call is for, or take a message for the individual.

Conference Room Phones and Video Systems in Teams

Microsoft Teams has the ability to connect speakerphones to the Teams environment so that a shared phone can be placed in the middle of a conference table and shared by others. Additionally, there are video phones that can be added to a conference room as well as full room size video conferencing systems that can be placed in a room for conference calling and information sharing.

Microsoft Teams can also connect to full room conferencing systems leveraging technologies like Microsoft's own Surface Hub device, similar to the one shown in Figure 6-1, that is a video conferencing and writable shared whiteboard solution as well through any of a number of third party provides like Polycom, Logitech, Crestron, or the like.

Figure 6-1: Microsoft Surface Hub Room Conferencing System

Additionally, Microsoft provides H.323 interoperability between Microsoft Teams conferences with other H.323 provides like Polycom, BlueJeans, and Pexip among others for system to system communications.

Making a Phone Call from Teams

Once a Cloud PBX license has been assigned to a user and the user has a dial plan associated to their account, when the user is in Microsoft Teams, a new icon appears on the left side of their Microsoft Teams interface that looks like a phone handset called "Calls". Clicking on the Calls button will open up a familiar dialpad page like the one shown in Figure 6-2 where the user can enter in the phone number and press the phone handset looking icon at the bottom middle of the dialpad to make the phone call.

Figure 6-2: Microsoft Team DialPad

When a phone call is established, the user can transfer the call or add other participants to the call by clicking on the "people+" icon, like the icon shown in Figure 6-3, in the upper right of the screen to add someone to the call.

Figure 6-3: People Icon to Call or Add Participants to a Conversation

Answering a Phone Call from Teams

Incoming calls will ring into the user's app and present a visual, audio, or both notifications to let the person know a call is coming in. The user can select to answer the call or can choose to not answer the call or have the call transferred to voicemail. This experience sometimes looks a little different between the Windows, Mac, iPhone, and Android apps, however the functionality is similar and the user will receive a call and have call controls to transfer or add additional participants to the call if desired.

Voicemail in Microsoft Teams

Part of Microsoft Teams is voicemail. When a call is received and the individual does not answer the phone, the caller has the option to leave a voicemail message. The message shows up in the Voicemail section in the Microsoft Teams client similar to what is shown in Figure 6-4. The user will be able to click on the voicemail and listen to their messages.

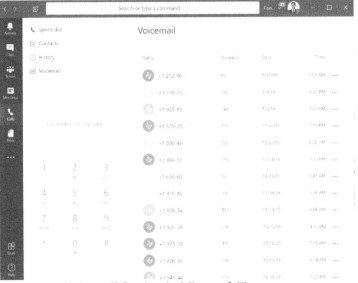

Figure 6-4: Voicemail Section in Microsoft Teams

Voicemails in Office 365 are stored in the cloud and can be listened to at any time. There are additional options when the user clicks on the "..." ellipse to the right of the voicemail message where the user can choose to:

- Mark the voicemail as unread
- Delete the voicemail
- Call the person back
- Add the person's phone number to speed dial
- Add the person to contacts
- Or Block the person

7 GETTING FAMILIAR WITH TEAMS AND LEVERAGING EVERY DAY BEST PRACTICES

There are a number of features and functions built in to Microsoft Teams that this chapter will highlight both from an awareness perspective, but also to share best practices how the functionality is leveraged every day by active Microsoft Team users.

Using the @mentions Function of Teams

One of the functions in Microsoft Teams is the ability to put an @ in front of a user's name in a conversation so that when the user enters Teams, they get an Activity notice that'll help them go straight to the conversations that call them out directly. As part of the day to day usage of Teams conversations, a user that steps away from Teams for a day due to a bunch of meetings, or is out for a couple days and is just getting back to work, trying to go through hours or days of conversations can be overwhelming.

However with @mentions prompts, the user heads straight to pertinent conversations they've been called out from. An example of this function:
@Rand, can you please comment on this topic about best practices for leaving messages to other team members?

Or possibly:
I know @Rand and @Chris are out of town this week, but when they get back, let's hear what they have to say

Jumping to @mentions Conversations

For the user who wants to jump straight to all of their @mentions, in the Team client, in the top "Search or type a command" box, the user can type in /mentions as shown in Figure 7-1.

Figure 7-1: Jumping to @mentions in Microsoft Teams

Alternately, the user can click on the "Activity" navigation item in the left side of the Teams client where the Activity option highlights mentions, replies, and other notifications for the user to key their attention to.

Using the / Command in Teams

The /mentions referenced in the previous section is one of many different / (slash) commands available in Microsoft Teams. Some of the other slash commands include:

- /activity: This allows a user to find the activity of another person in the organization or on their team, using it such as /activity *Andrew* will show Andrew's recent activity in Teams.
- /available, /away, /brb, /busy, /dnd are all options to set a user's status or availability. Obviously /available will set the user as available whereas /away notes the person has stepped away from their desk. The /brb (be right back) or /dnd (do not disturb) provide an timing awareness of whether the person is available for a chat in Teams.
- /call, /chat will quickly call or start a chat with someone else in the organization or on your team. This is used like /call *Rich* will start a call to Rich in the organization
- /files provides a list of the most recent files that the user has access, which is helpful if the user recently opened a file and wants

to go back and open up the same file and eliminates the user's need to go search for the file through a series of teams and channels.

- /goto sends a user to a specific team or channel that the person names, such as /goto *All Staff* channel.
- /join quickly enables a user to join a team
- /org pulls up the organization's org chart
- /saved, /unread prompts the user for any saved items or activities that have been unread, helping the user find things more quickly
- /whatsnew provides a user a quick view of new things in Microsoft Teams. Since Teams comes out with new features every few weeks, this is helpful to glance at every now and then to see if there's something new that might be of interest
- /help provides information about Help topics for Microsoft Teams
- /keys provides a list of keyboard shortcuts, that will be covered in more detail in the next section.

Keyboard Shortcuts in Microsoft Teams

As much as Microsoft Teams has slash commands that help a user quickly navigate around and perform quick tasks like looking up files, starting chats and calls, or looking for @mentions, Teams also has a series of keyboard shortcuts for those good with using Ctrl-key (Windows) or Command-key (Apple Mac) shortcuts.

The common (general) Windows-based keyboard shortcuts include:
- Ctrl . (ctrl-period): Shows keyboard shortcuts
- Ctrl / (ctrl-slash): Shows commands
- Ctrl N: Starts a new chat
- F1: Opens help
- Ctrl = (ctrl-equal sign): Zooms in
- Ctrl – (ctrl-minus sign): Zoom out
- Ctrl 0 (ctrl-zero): Resets Zoom level
- Ctrl E: Go to search
- Ctrl G: Goto
- Ctrl , (ctrl-comma): Open Settings
- Esc (escape key): Close

The common (navigation) Windows-based keyboard shortcuts include:
- Ctrl 1: Open Activity
- Ctrl 2: Open Chat
- Ctrl 3: Open Teams
- Ctrl 4: Open Meetings
- Ctrl 5: Open Calls
- Ctrl 6: Open Files
- Alt up (Alt-UpArrow): Go to previous list item
- Alt down (Alt-DownArrow): Go to next list item

The common (messaging) Windows-based keyboard shortcuts include:
- C: Go to compose box
- Ctrl-Shift-X: Expand compose box
- Ctrl-Enter: Send (expanded compose box)
- Ctrl-O: Attach a file
- Shift-Enter: Start a new line
- R: Reply to a thread

The common (meetings and calls) Windows-based keyboard shortcuts include:
- Ctrl-Shift-A: Accept a video call
- Ctrl-Shift-S: Accept an audio call
- Ctrl-Shift-D: Decline a call
- Ctrl-Shift-C: Start an audio call
- Ctrl-Shift-U: Start a video call
- Ctrl-Shift-M: Mute a call
- Ctrl-Shift-E: Start screen share session
- Ctrl-Shift-O: Toggle video
- Ctrl-Shift-F: Toggle full screen
- Ctrl-Shift-Space: Go to sharing toolbar
- Ctrl-Shift-A: Accept screen share
- Ctrl-Shift-D: Decline screen share
- Ctrl-Shift-P: Toggle background blue

For all of these keyboard shortcuts, the function on an Apple Mac system is to use the Apple Command key instead of the Ctrl key, so a Ctrl-1 on Windows is Command-1 on a Mac. Or a Ctrl-Shift-X on a Windows system is Command-Shift-X on a Mac.

Adding Tabs to a Team and Channel

Microsoft teams and channels have various tabs that expand more groups of information for the team or channel users to access. By default, there is a Conversations tab and usually a Files tab, and then after that it is up to the users of the team or channel to select other tabs.

There is a + button on the tab that allows the addition of tabs. Then a user presses the + button, a list of optional tabs to be added shows up, as shown in Figure 7-2

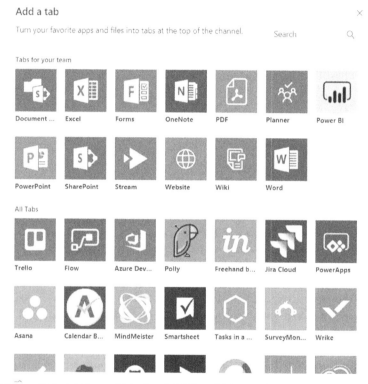

Figure 7-2: Adding a Tab for Additional Teams Services

There are a number of Microsoft Office 365 component tabs that can be added that include things like:

- Planner: To add a tab for a Microsoft Planner worksheet
- OneNote: To add a tab for a shared OneNote document
- Stream: To add a tab for shared videos to be available
- Wiki: To add a tab for a shared Wiki that the team might use to

input in notes and ideas
- PDF: To add a tab for a specific PDF file to be linked and openable when the tab is clicked
- Website: To add a link to a website that the users of the team or channel want to go to when the tab is clicked.
- Etc

In addition to Microsoft Office 365 component tabs, Microsoft also provides integration with 3rd party services that can be accessed when users click on a tab. Some of the services available when a tab is clicked:

- Trello: Provides boards, lists, and charts
- Polly: A polling app to create polls and surveys within Teams
- MeisterTask: A task management and collaboration add-in
- Quizlet: A learning tool that allows Teams members to create flashcards, games, and learning tools to Teams
- Asana: a task assignment and due date tracking app
- Etc

Many different add-ins that can have a tab added to teams and channels to help organizations find, organize, track, manage, and communicate more efficiently and effectively.

Using the Wiki Function in Teams to Roll-up Information

Microsoft Teams provides a persistent chat that users can carry on conversations and share content with one another, however for a new participant to jump into weeks of conversations, it becomes time consuming and many times confusing for the person to get up to speed on the latest conversations. One of the best practices leveraged is utilizing the Wiki feature in Microsoft Teams where a content facilitator for the team or channel will roll-up the latest information into a Wiki. This roll-up of information, similar to what is shown in Figure 7-3, helps to keep pertinent information organized, shared, and understood by groups of users.

Figure 7-3: Wiki Roll-up of Teams Conversations

Sharing Group Calendars

This is something that SharePoint and pretty much all other content sharing, group chat, and collaboration tools struggle with in handing group calendars that Teams integrates nicely (straight in Microsoft Outlook!)

Each team, group, or project can have their own calendar with deadlines, milestones, due dates, etc that the project or team is driving toward. Instead of polluting a user's personal calendar with a bunch of project notes, or having a completely separate calendar that requires a user to "toggle" between their personal Outlook calendar and an external calendar, within Microsoft Outlook, a user can pull up their Teams calendars and view them right along their personal calendar similar to what is shown in Figure 7-4.

Figure 7-4: Viewing Personal and Team/Group Calendars in Outlook

With a simple "click" at the top of the calendar, a user can turn on/off the views of the various group and personal calendars they want to view, compare, and access.

Adding in Microsoft Planner and Stream to Teams

As mentioned earlier, Microsoft teams and channels can have other Microsoft components (+) added in as tabs to Teams. Microsoft Planner handles project tasks and Microsoft Stream manages shared video content. Both of these tools are common add-ins used in Microsoft Teams.

Microsoft Streams, shown in Figure 7-5, is Microsoft's version of YouTube videos that allows organizations to upload videos to be shared with the Team. It might be a training video, could be a video capture of a Web meeting. Unlike YouTube or other external video services, Streams ties directly into Office 365 security, so the content can only be accessed by users with the appropriate permissions. Streams also does audio to text conversion so that within minutes of uploading a video, the content of the video is transcribed for quick and easy search and access.

Figure 7-5: Microsoft Stream Video Sharing in Office 365

Microsoft Planner is a lite project management tool that allows organizations to enter in tasks, milestones, create dependencies, assign users, assign deadlines, and track projects and initiatives. Unlike the full Microsoft Project program that has a lot of complexities to it, Planner is a very simplified web-based solution, shown in Figure 7-6, that any user who can click to add a task and type in a handful of fields can create tasks and manage tasks for a team.

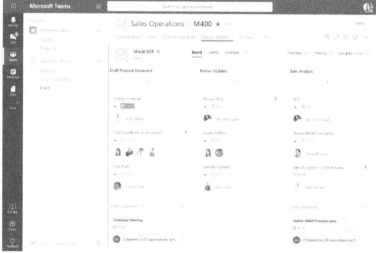

Figure 7-6: Microsoft Planner Light Project Coordination and Management

Customized Landing Page

Some organizations, while leverage the full functionality of Microsoft Teams for file sharing, collaboration, calendars, etc, still prefer to have a Landing Page where users in various departments can go to for centralized information.

This can be done within the Microsoft Teams construct without having to build a completely separate Intranet site or series of pages. These landing pages, like the one shown in Figure 7-7, are built from Microsoft SharePoint templates and can be linked and accessed by users of the team.

Figure 7-7: Sample Landing Page in SharePoint Tied to Teams

To create a SharePoint landing page for Microsoft Teams, from within Office 365 Outlook Web (or in the Teams app) where the Conversations, Files, Calendar tabs are located is an (ellipse) . . . Click on the ellipse and select "Site" that'll bring the user to a page for the Team site similar to the one below. The Team site (web view) is a fully editable SharePoint landing page that an organization can customize the colors, graphics, web part sections, etc to be more "web-like" for users. Organizations can link different document libraries, 3rd party SaaS app links, etc to this page. This page is created by default when a team site is created and is directly linked to the Team site for the users to go to.

At any point if the user wants to switch back to Outlook / Groups, they simply click on the "Conversations" option in the left column of the Site page and that'll bring the user back to the Outlook Web view. This allows a user to move back and forth between a customized Web/Intranet type view and Outlook/Groups.

8 MIGRATING FROM SLACK TO TEAMS

A common query by organizations is to understand how Microsoft Teams compares with Slack. Slack is typically already embedded in organizations and the decision is whether Microsoft Teams can do the same thing that Slack does, and what's involved in migrating Slack content to Teams if the organization chooses to standardize on Teams.

Teams vs Slack Functionality

Having used Slack for years and gone through the process of helping dozens of organizations shift from Slack to Teams, end of the day the products do the EXACT SAME THING. In some cases they are same, but operate differently, but in a lot of functionality they are actually identical.

The two biggest complaints from Slack users is having to learn and use a new tool (and one from "Microsoft" at that) but the bigger complaint is migrating content from Slack to Teams.

For the first piece of learning something new, no one wants to figure out a new tool and learn something different, so the best way to get over the complaint of users having to learn something new is simply have it mandated by someone high enough in the organization to just say that "we're migrating to Teams." Without executive mandate, Slack users will complain, and that's either a stopper that will prevent the organization from every migrating to Teams, of just part of the normal process of "change."

As for the challenge of migrating existing Slack content to Teams, that is addressed later in this chapter. There are tools to migrate content as well as best practices in simplifying the process of migrating "everything" from Slack to Teams. Look at the section later in this chapter that addresses content migration for details.

Beyond these two complaints, Teams and Slack do the same thing. There's a conversation / chat area where users can type in information, and reply and respond to others.

Slack is a flat hierarchy with "channels" that can be set to public (anyone can access) or private (limited on access). Microsoft has a two layer structure where "teams" are created, and then "channels" are created under the teams.

In Microsoft Teams, users are assigned to the team, and the team can be made public (viewable by others) and private (not viewable by others). Slack users will say the "problem with Teams is that you cannot make a Teams channel private", which is true, all channels in Teams nest under a team. However if you really want a conversation that is private and just between a limited number of users, then simply create a new team and assign users to that team and flag the team as private. That is doing the EXACT same thing as Slack. In Slack, you create a channel, add users to the channel, and flag the channel as private. In Teams, you create a team, add users to the team, and flag the team as private.

What is nice about Microsoft Teams that Slack users eventually appreciate is the ability of creating a team once and simply adding multiple channels under the team. The team can be a department, an office site, or some other user designation. The channels in Microsoft Teams spawn off conversation topics, so a channel might be about a product the team is working on, or another channel might be about a marketing campaign being undertaken by a group. The team is the security and user boundary, and the channels are subtopics within that team.

From basic functionality, both Slack and Teams have @mentions and slash (/) commands as well as shortcuts. Both Slack and Teams provide users the ability to post files, collaborate, and communicate.

3rd Party Integration and Support

Organizations that use Slack typically integrate Slack with 3rd party providers for file sharing (like Google Docs, Box.com, or the like), integrate with Web and Video conferencing solutions (like Zoom, WebEx, BlueJeans, etc) and integrate with security and compliance add-ins.

Microsoft Teams is fully integrated into Microsoft Office 365 that comes with the Microsoft's email system, SharePoint and OneDrive file collaboration and sharing, Teams/Skype Web conferencing, Video conferencing, and Cloud PBX telephony, as well as complete integration of security components for legal hold, eDiscovery, archiving, and content management.

This is an area that again is the same between the two platforms, but different. Users who want to keep Google docs, BlueJeans, and Symantec security that they have tied in integrations with 3rd party products will find that while you CAN integrate 3rd party products to Microsoft Teams, the question is why would you?

The benefit Microsoft brings is the ability to get rid of a dozen one-off products and bring the functionality, security, collaboration, unified licensing, and support into a single simplified platform.

However there are a number of 3rd party integrations available should organizations truly want to integrate with Google docs (that can be done) or with BlueJeans (that can be done), as other 3rd party integrations like Jira Cloud, Survey Monkey, Twitter, etc.

The best practice in considering migrating 3rd party integration components to Microsoft Teams is to determine whether the functionality is native to one of Microsoft's Office 365 components or whether the 3rd party component is truly unique. Anything that can leverage existing technologies in Microsoft's Office 365 suite provides one more thing an organization can get rid of and simplify their collaboration and communications environment.

For functionality that might not exist in Office 365 natively and not available in 3rd party components, there's always the ability to leverage the APIs and programmatic functionality in Teams for unique customization.

Microsoft Teams has a public API that provides access to programmatically query, process, write, and control attributes and data within Teams. Microsoft uses what is called Graph API. Graph API provides controls to teams, groups, channels, apps, chat, and calls with various functions to add, remove, update, delete, or post data, fields, content, or information.

More details about Graph API for Teams is at https://docs.microsoft.com/en-us/graph/api/resources/teams-api-overview

Slack User Adoption of Microsoft Teams

User adoption by Slack users to Microsoft Teams is actually a LOT easier process of getting users to get familiar and adopt Teams compared to those who have never used a content collaboration or communications tool in the past.

Because Slack users know the fundamental functions of channels, file folders, content sharing, and collaboration, they merely need to relearn how Teams does it versus how they did it in Slack. Much of the content in this book from start to finish has been around helping users understand things like using Teams for chat, conferencing, calls, collaboration, and secured communications.

Migrating Slack Content to Teams

As noted earlier in this chapter, one of the common requests of Slack users is the ability to seamlessly migrate Slack content to Microsoft Teams. Content migration CAN be done, however experience has shown that once Slack users know and see what all Microsoft Teams does, fewer than 15% of those organization have 100% of their Slack content migrated to Teams.

However, for organizations that do want/need their historical content migrated from Slack to Teams, by leveraging the Slack API, organizations can export content out of Slack and then import that content into Microsoft Teams.

A very simple way to get information across from Slack to Teams is to export Slack content and simply import the content and conversations into a Word or Wiki file for the new Microsoft team and channel. This "document" for each Slack channel provides a searchable way for users to access old Slack content, but allows the organization to start from scratch and rethink how Teams can support the management, access, security, control, and usability of information better than what was originally created in Slack.

When organizations spend a lot of time trying to make Microsoft Teams into "exactly" what Slack was, they completely miss the functionality native to Teams. The migration centric organization tries to integrate the BlueJeans 3rd party plug-in to Teams as opposed to just using the native Microsoft Teams Web, Video, and Audio conferencing functionality.

Again, if you want to make Teams a mirror of what Slack was, that CAN be done with content export and import, however rethinking, re-engineering, and improving processes, workflow, communications, and information management can help organizations work smarter and better, than replicating an older process.

As for selecting content to be migrated from Slack to Teams, the other thing an organization needs to take in account is how much of the information is historical and can merely be migrated over as archived content and doesn't need to be put into active Microsoft Teams channels.

Most organizations that have been using Slack for a long time find that 70-80% of the content is old and unused. As much as the content could be migrated to Teams, a lot of that content can simply be deleted as being irrelevant, or at most just exported to a searchable database or file for future look-up at best.

Organizations use the Slack API to identify the age of content and last access of the information to determine whether content needs to be migrated to an active channel.

And for many organizations, having a shift from Slack to Teams that is done over a 2-3 month period will find that as they have users spin up NEW content in Microsoft Teams as teams or channels, and continue to use Slack for a short period of time during the transition, that after a couple months, most active communication and collaboration shares are being done in Microsoft Teams and the use of Slack, even for the largest enterprises is relegated down to just 30 or 50 Slack channels which is a lot more manageable to migrate and convert than what most organizations start off thinking that they have "thousands" of Slack channels that are "used every single day by users" and that there's "no way the organization can shift the platform to a new system."

It's amazing that by identifying exactly what Slack is being used for, educate users how Teams works and what can be done with Teams, and with a little nudge from executive management, organizations shift from Slack to Teams relatively easily.

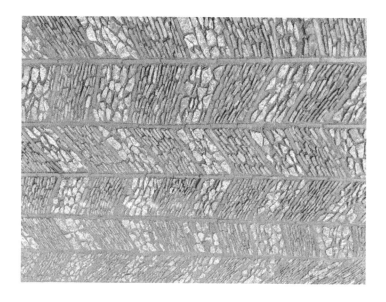

9 ADMINISTERING AND MANAGING TEAMS

As organizations adopt Microsoft Teams, it becomes important for I.T. to properly enable security and manage content to ensure the various intelligent communication mediums are properly managed and administered.

While Chapter 2 "Acquiring and Enabling Microsoft Teams" covered the basics of enabling users to access Microsoft Teams, this chapter goes into more organizational settings focused on administering and managing Teams.

Accessing the Microsoft Teams Admin Console

The Microsoft Teams Admin Console is accessed from https://portal.office.com, click on the "Admin Centers" in the bottom left navigation bar, and select "Teams & Skype" as shown in Figure 9-1.

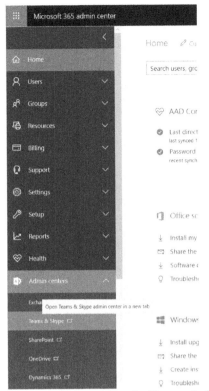

Figure 9-1: Accessing the Teams & Skype Admin Center

The Microsoft Teams & Skype for Business Admin Center provides a number of different navigation selections to configure settings from meeting defaults, Cloud PBX configuration settings, call quality dashboard views, and reporting components.

Enabling External Communications with Teams Users

One of the first settings organizations should review and set are those that are tied to the "Org-wide Settings" options in the Microsoft Teams & Skype for Business Admin Center. The setting to enable "External Access" is a decision the organization needs to make.

By default, organizations leave this "External Access" setting to ON. This allows users from Teams to Instant Message and Chat with people outside the organization. Only in rare instances does an organization want Teams to only be an internal communications tool. It's like having email that only sends and receives messages to internal employees, and emails never being sent or received outside of the organization.

However in a handful of cases such as in a case where an organization may want emails to be internal and external, but Teams to be internal only, then having the organization set this "External Access" setting to OFF makes sense.

Enabling External Access to Teams Content

An extension to the "External Access" setting is the "Guess Access" setting. Where the "External Access" setting is for Instant Messaging and Chat, this "Guest Access" is specific to sharing files and content within Teams to users outside of the organization.

This gives an organization the ability to Chat with people outside the organization, but potentially not allow the internal users to share content with users outside of the organization. If the organization chooses to not allow content to be shared outside the organization, then the "Allow guest access in Microsoft Teams" should be set to OFF.

Additional granular settings in the "Guest Access" section shown in Figure 9-2 controls external guest participants the ability to make calls, participate in video sessions, share their screen, or use the "Meet Now" feature of starting a Web meeting. So an organization can allow external guest access, but limit some of the granular functions the user can do.

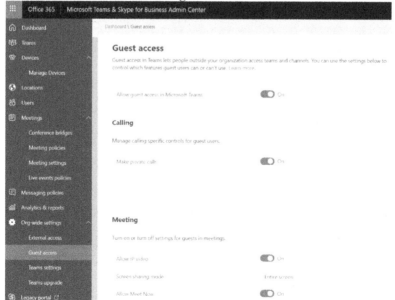

Figure 9-2: Guest Access Options in the Teams & Skype Admin Center

Team Settings in the Admin Center

Additional settings are available under the "Team Settings" within the Microsoft Teams & Skype Admin Center that allow administrators to control whether users can send emails using the channel email address, and whether users can link an external content sharing platform (like Box, ShareFile, DropBox, and Google Drive) to a Teams channel.

Other "Team Settings" include device control for meeting room devices, whether secondary authentication is required to join meetings from a meeting room and whether the resource accounts used for meeting rooms can send messages on behalf of the room or resource. These are granular settings that help administrators limit access where appropriate.

There is no specific "best practice" on settings, however the defaults are usually what organizations have set, and unless there are specific reasons the organization would change the defaults, leaving them "as is" is most common for enterprises. When in doubt, think about how the organization handles email settings as well as if the organization had any specific settings in Box, Google Docs, OneDrive or the like that the organization may choose to mirror prior settings.

Location and Devices in the Admin Center

Within the Microsoft Teams & Skype Admin Center are settings that allow administrators to control device locations and identify devices connected to the Teams environment.

Under the Locations navigation bar setting in the Admin Center, administrators can create location or site designations and assign IP address identifiers in Teams to denote the site. For Microsoft teams that spans multiple sites, there are configuration settings that identify and allow access from certain sites, or identifies dial plan routing of communications between locations based on site locality.

Under the Devices navigation bar setting in the Admin Center, administrators can enter the devices that are connecting to Teams such as room size video conferencing systems, or table top conference phones. By identifying devices, a device policy can be placed on the devices that enables configuration settings for the devices.

Appliance devices many times do not have extensive device configuration consoles, so assigning an IP address versus using DHCP and dynamically assigning an IP address to the device can be important in device configuration and administration.

A device policy can be set require a PIN to unlock a shared conference room device so that not everyone can walk into a conference room and start making long distance phone calls or anonymously joining meetings.

Configuring Teams Meeting Settings

When users are invited to Microsoft Teams meetings, the default invitation verbiage is something that the organization can customize. The invite can include the logo for the organization, any specific legal wording or terms of use specific to joining a Web meeting.

The administrator can go into the "Meetings / Meeting Settings" option in the Admin Center to see what is shown in Figure 9-3 to get to the section that allows configuration settings for the meeting invite verbiage.

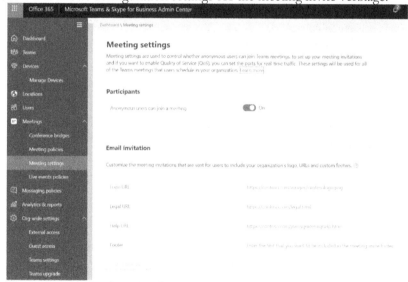

Figure 9-3: Teams Meeting Setting Options

Generating Usage Reports in Microsoft Teams

As with any application, Microsoft Teams provides usage reports that can provide administrators in the organization insight on user access and usage of Microsoft Teams.

The administrator can go to the "Analytics & Reports" navigation option in the Admin Center and there are options for the administrator to dump reports including:

- Teams User Activity: Teams User Activity shows the users who are most active in Microsoft Teams, the # of channel messages, chat messages, and 1:1 calls for each user.
- Teams Device Usage: Team Device Usage shows the make-up of devices accessing Teams, whether it's Windows, Mac, iOS, or Android devices connected and interacting with Teams.
- Teams Usage: Teams Usage, similar to one shown in Figure 9-4, shows which teams and channels are used and accessed most frequently. The report shows active users, # of external guests, and the active channels within each Team

Figure 9-4: Teams Analytics & Reports - Teams Usage Report

Leveraging Content Classification, Protections, and Performing eDiscovery Search and Legal Hold on Content in Teams

In working with Microsoft Teams, organizations may choose to classify content to prevent data leakage, or manage content in a more secure manner to address content retention and eDiscovery search.

Microsoft Teams content falls under Microsoft's security and compliance management structure that provides extensive controls for content protection. Content can be tagged and encrypted to prevent content from being deleted from the organization to address legal hold requests and compliance requirements. Content can also be tagged and encrypted so that users outside the organization cannot access content because they lack valid credentials to open protected content.

The various options available for organizations around security and content protections available in Microsoft Teams can be reviewed with detailed implementation instructions from the book "Handling Electronically Stored Information in the Era of the Cloud," that can be downloaded in PDF or Kindle format from http://www.cco.com/our-publications.htm

ABOUT THE AUTHORS

Rand Morimoto, Ph.D.: Rand has a unique blend of deep technical knowledge and expertise, and an academic background in organizational behavior and organizational management.

Rand blends the theory of business and economics with his knowledge and day to day experience in the tech industry, resulting in the content highlighted in this book.

David Ross, MCITP: David has over 20 years experience in I.T. consulting, the majority of which has been spent playing the lead architect role on network design and implementation projects throughout the San Francisco Bay area.

David is currently a principal consultant for Convergent Computing, and acts as practice lead for unified communications (U.C.) technology, assisting organizations of all sizes to meet their U.C. requirements leveraging Microsoft Teams, Skype for Business, and Skype for Business Online.

Made in the USA
Monee, IL
21 March 2020